My First Look at BODY SYSTEMS

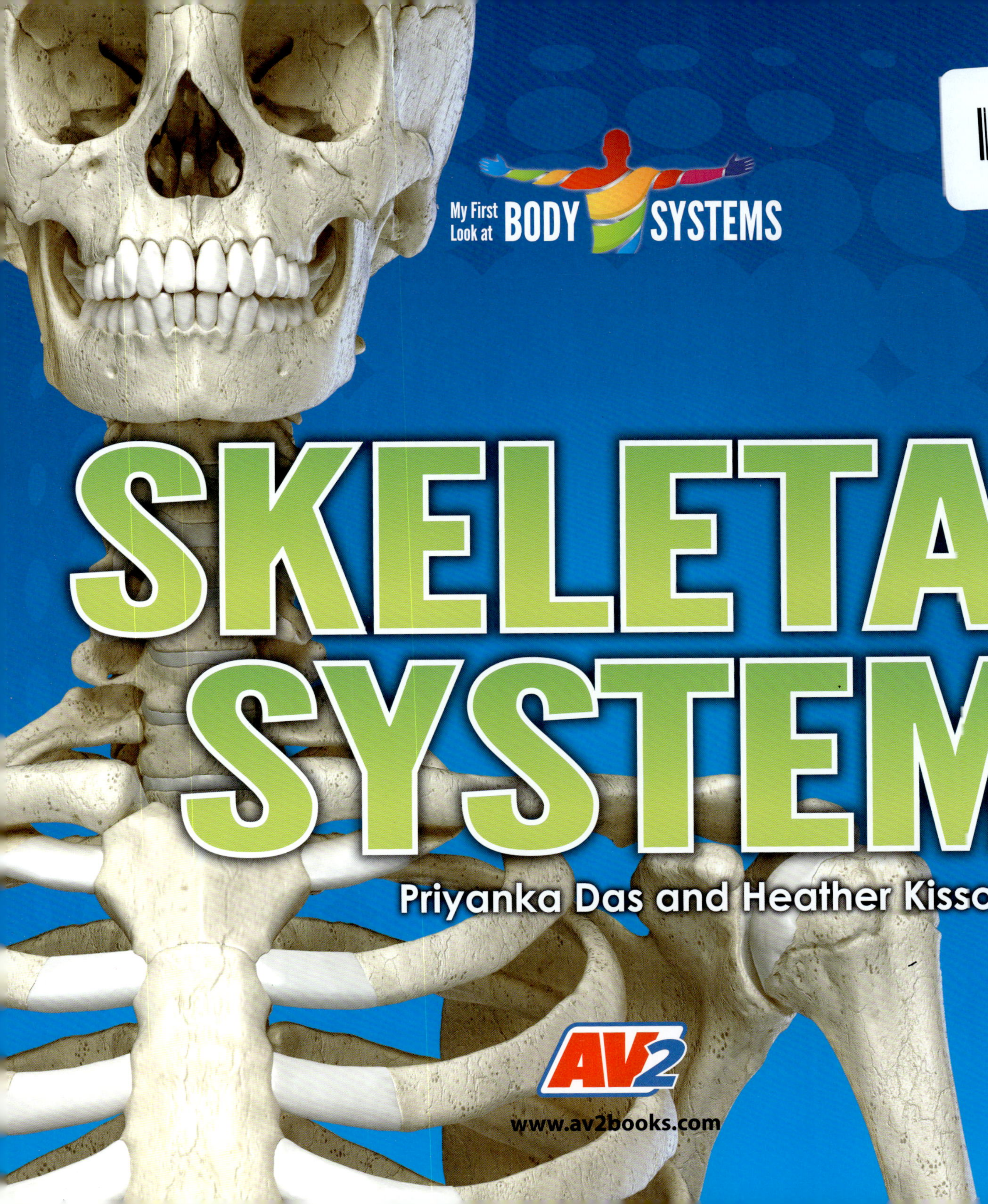

SKELETAL SYSTEM

Priyanka Das and Heather Kissock

AV2
www.av2books.com

Step 1
Go to **www.av2books.com**

Step 2
Enter this unique code

YVRDN7TQ3

Step 3
Explore your interactive eBook!

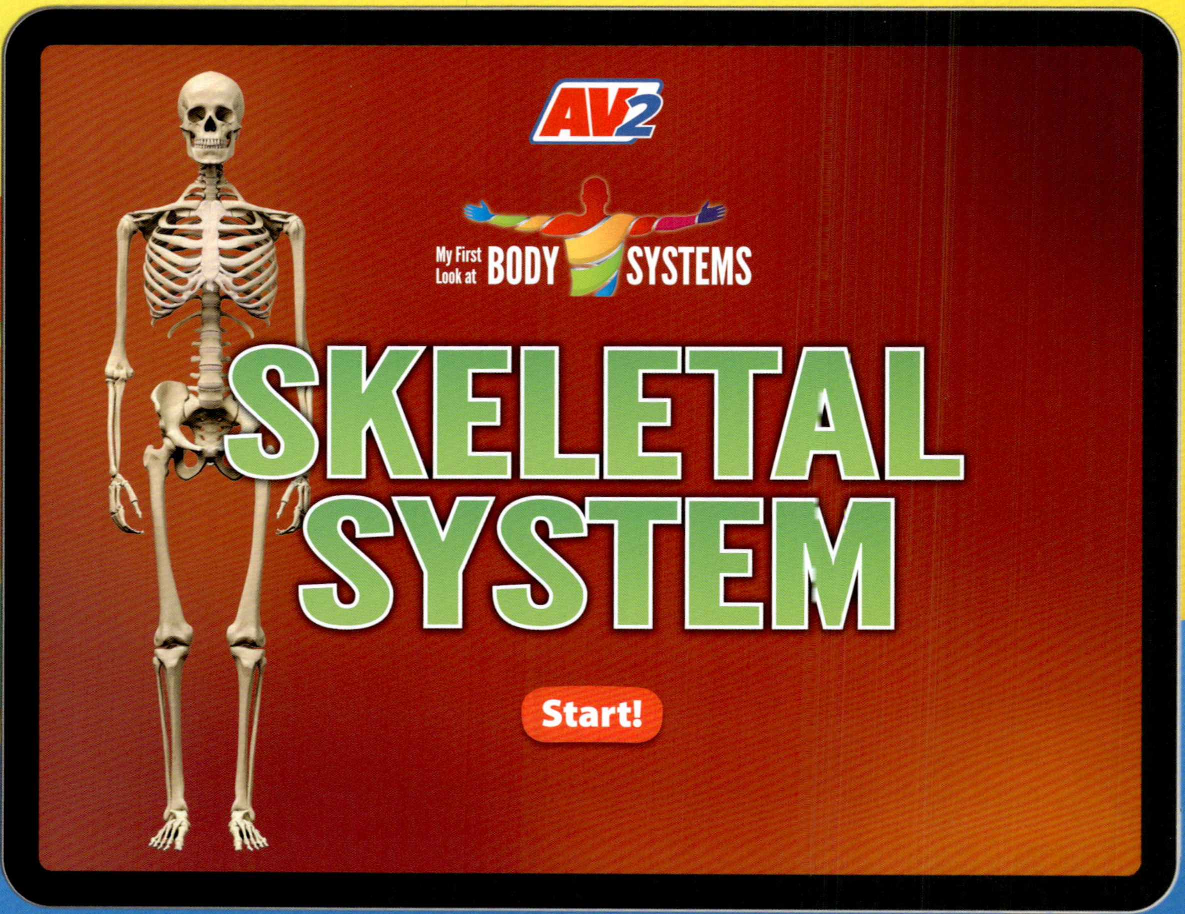

AV2 is optimized for use on any device

Your interactive eBook comes with...

Read
Audio
Listen to the entire book read aloud

Videos
Watch informative video clips

Weblinks
Gain additional information for research

Try This!
Complete activities and hands-on experiments

Key Words
Study vocabulary, and complete a matching word activity

Quizzes
Test your knowledge

Slideshows
View images and captions

View new titles and product videos at www.av2books.com

SKELETAL SYSTEM

My First Look at BODY SYSTEMS

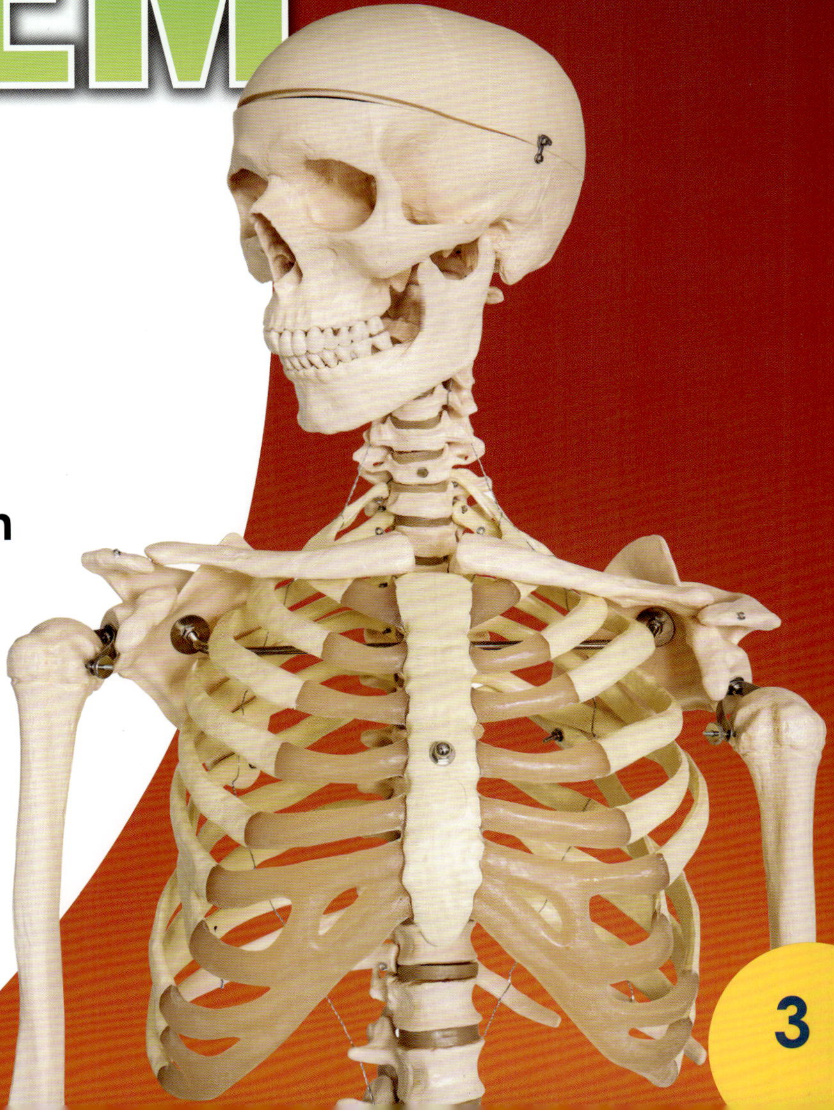

CONTENTS

Skeletal System

Think of your body as a powerful machine. Like all machines, it is made of systems. Each system performs an important job. **The skeletal system supports the rest of the body.** It gives the body its shape.

The bones of the skeleton are incredibly strong. **They are about five times stronger than steel.** Bones protect important organs, such as the heart and brain, from harm.

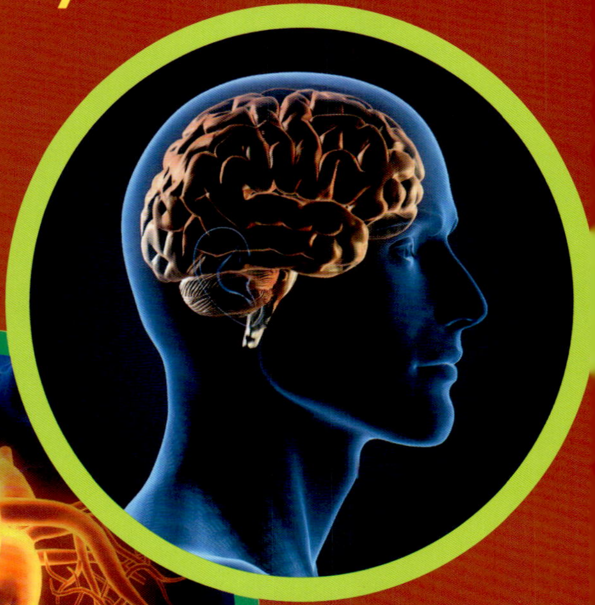

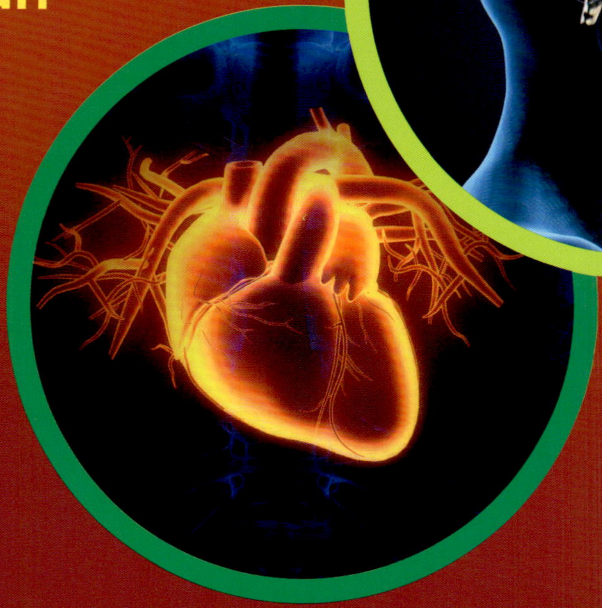

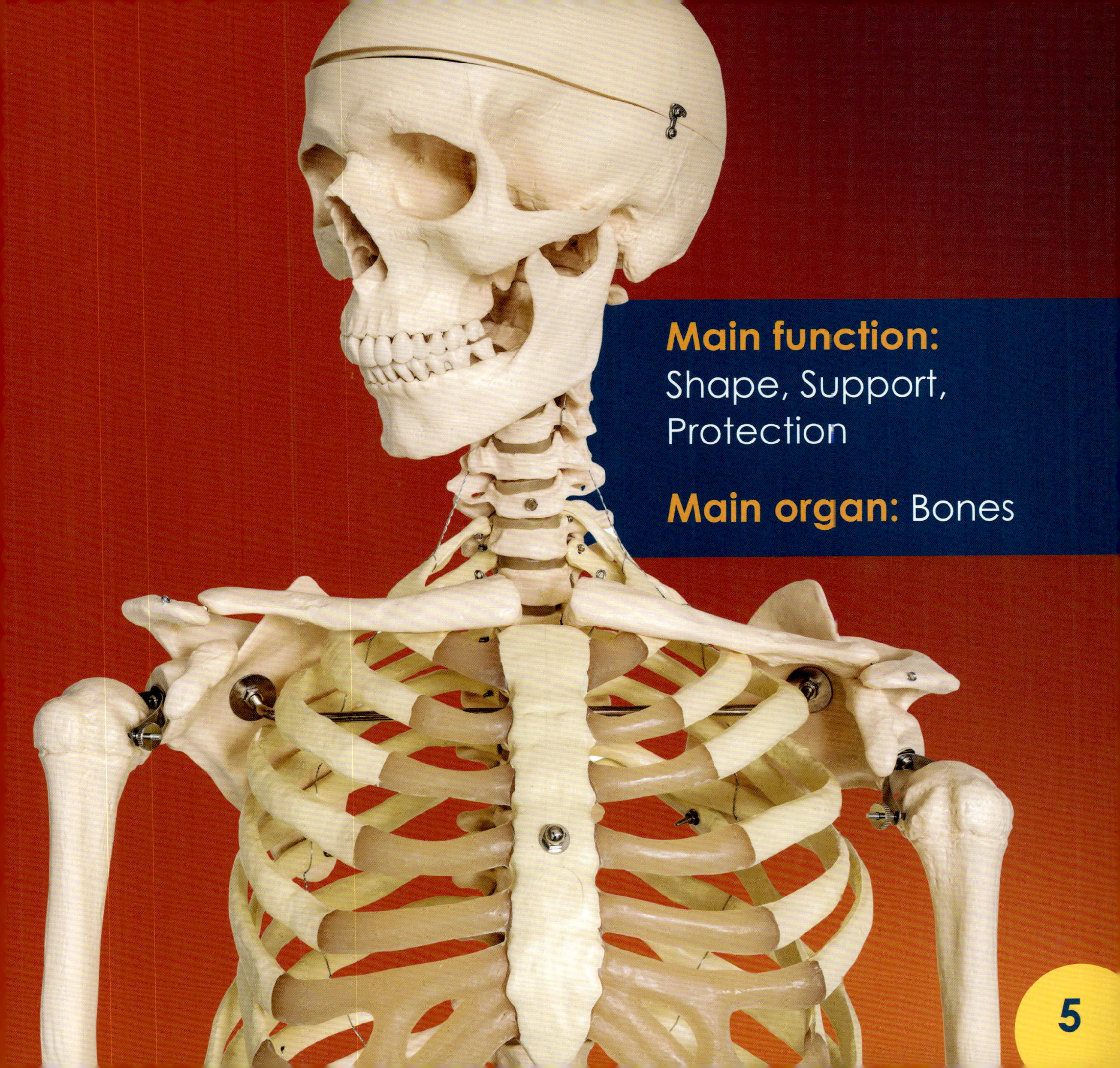

Main function: Shape, Support, Protection

Main organ: Bones

All about Bones

Bones are living, breathing organs. Like the rest of the body, they grow. **Blood carries oxygen to bones.**

If a bone breaks, it can repair itself. **Bones take a few weeks to heal.** Children's bones heal much faster than adults' bones.

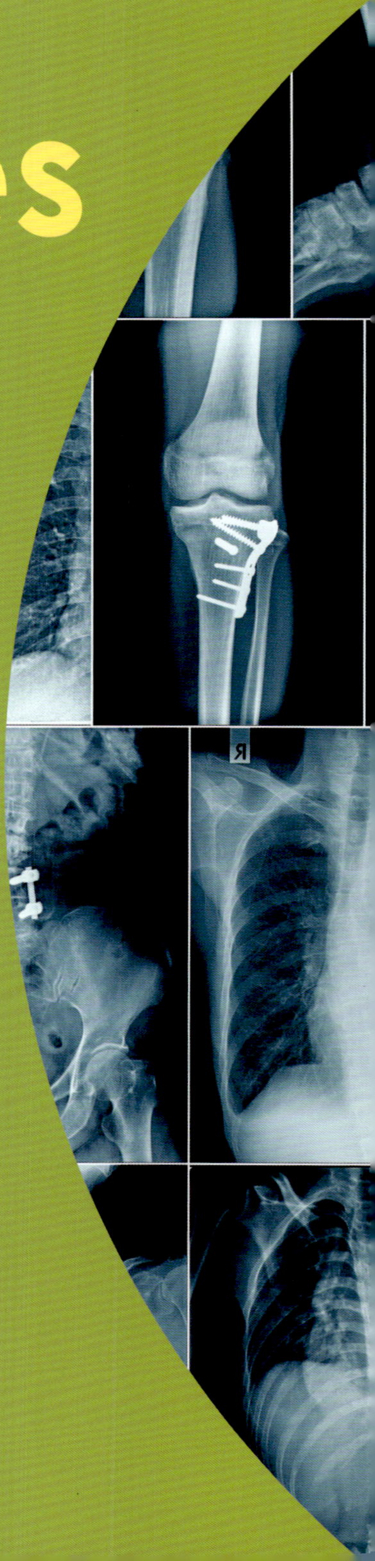

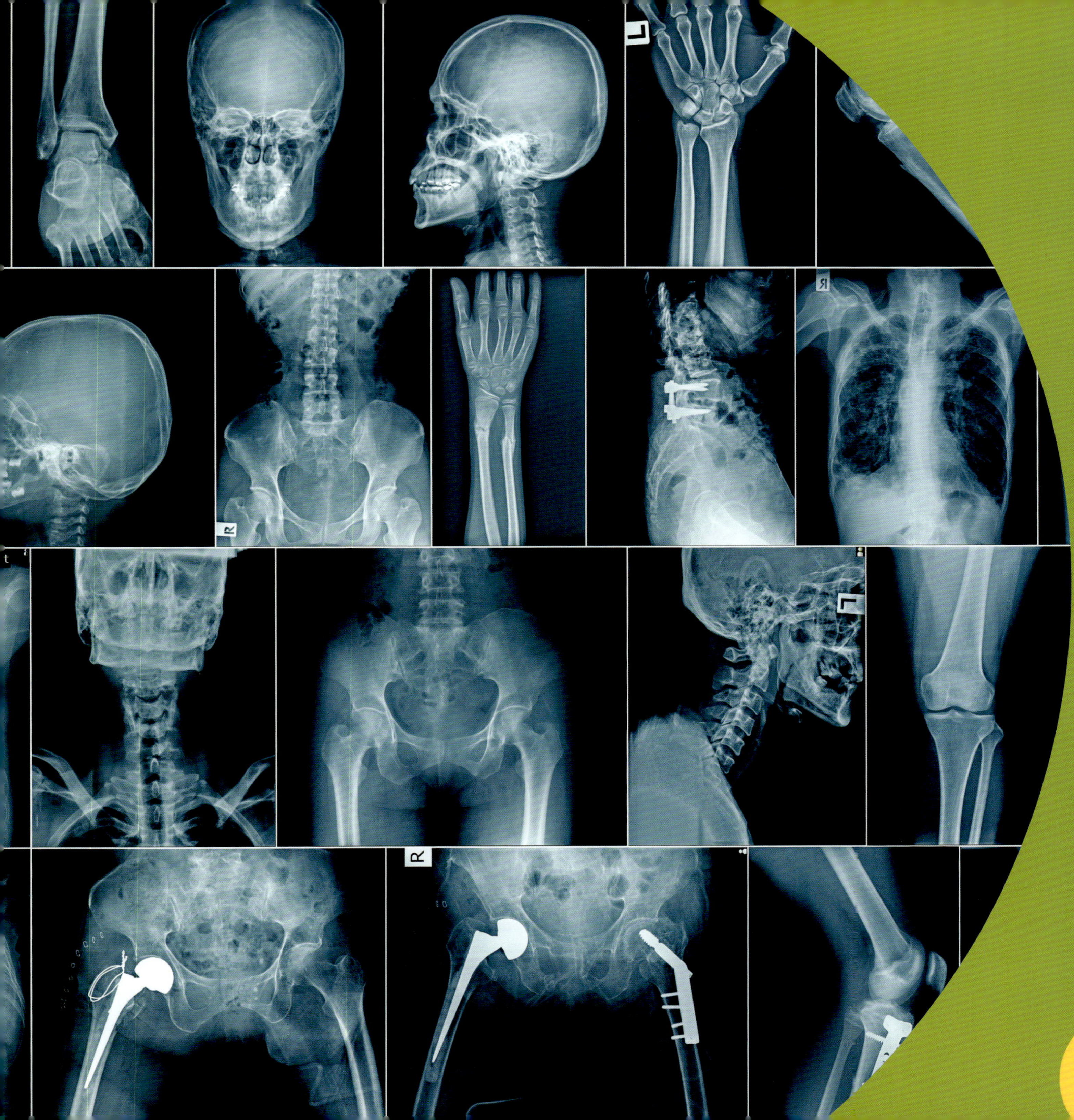

Parts of the Skeletal System

Babies are born with about 300 bones. These join together as children grow. **Adults have 206 bones.**

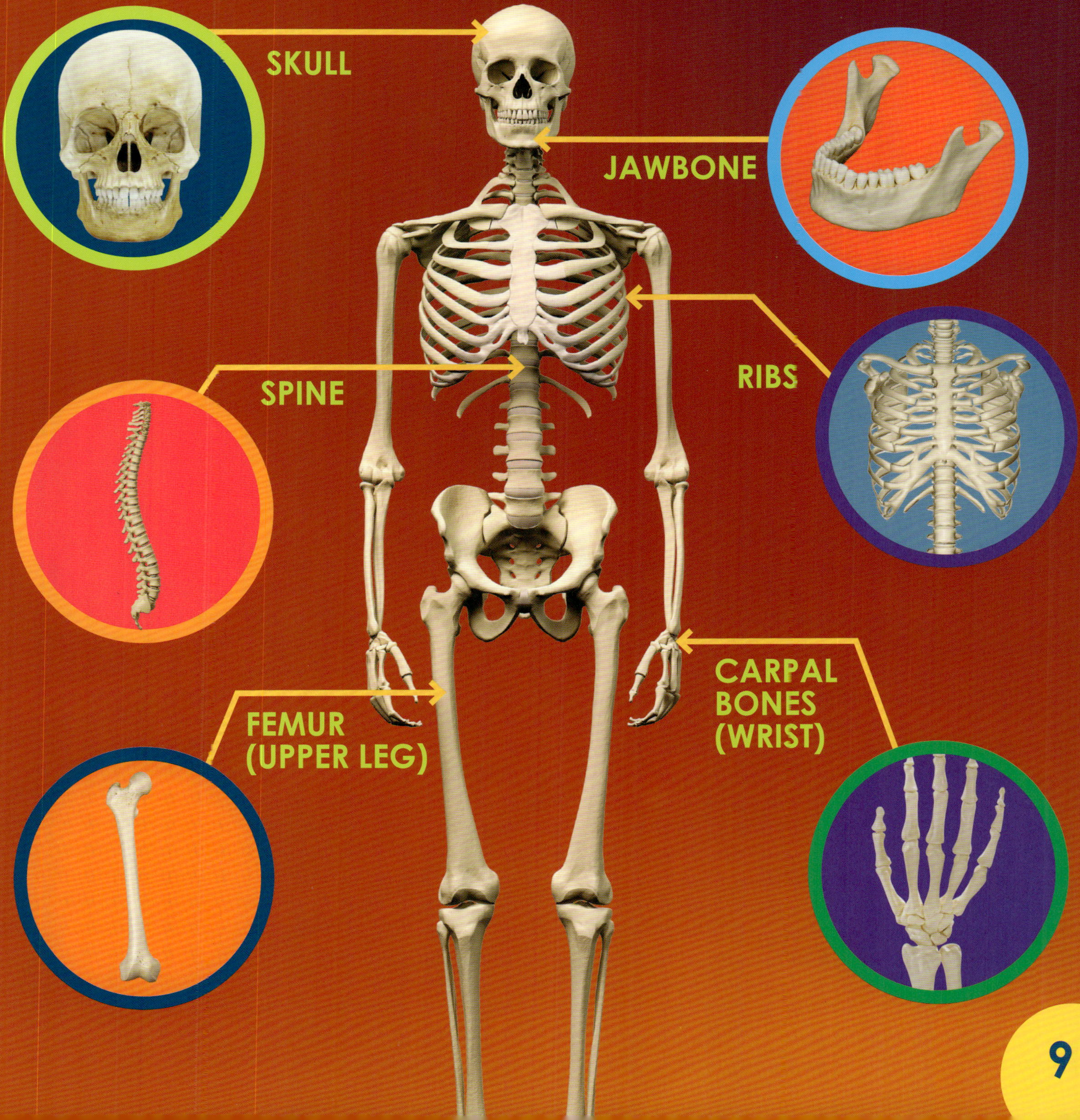
SKULL
JAWBONE
RIBS
SPINE
CARPAL BONES (WRIST)
FEMUR (UPPER LEG)

Skull

The skull shapes the face and protects the brain. Its bones support the eyes, nose, and mouth. Teeth are part of the skull, too. However, they are not bones.

There is only one movable bone in the skull. It is called the jawbone. The jawbone is at the bottom of the face. The other 21 bones in the skull do not move.

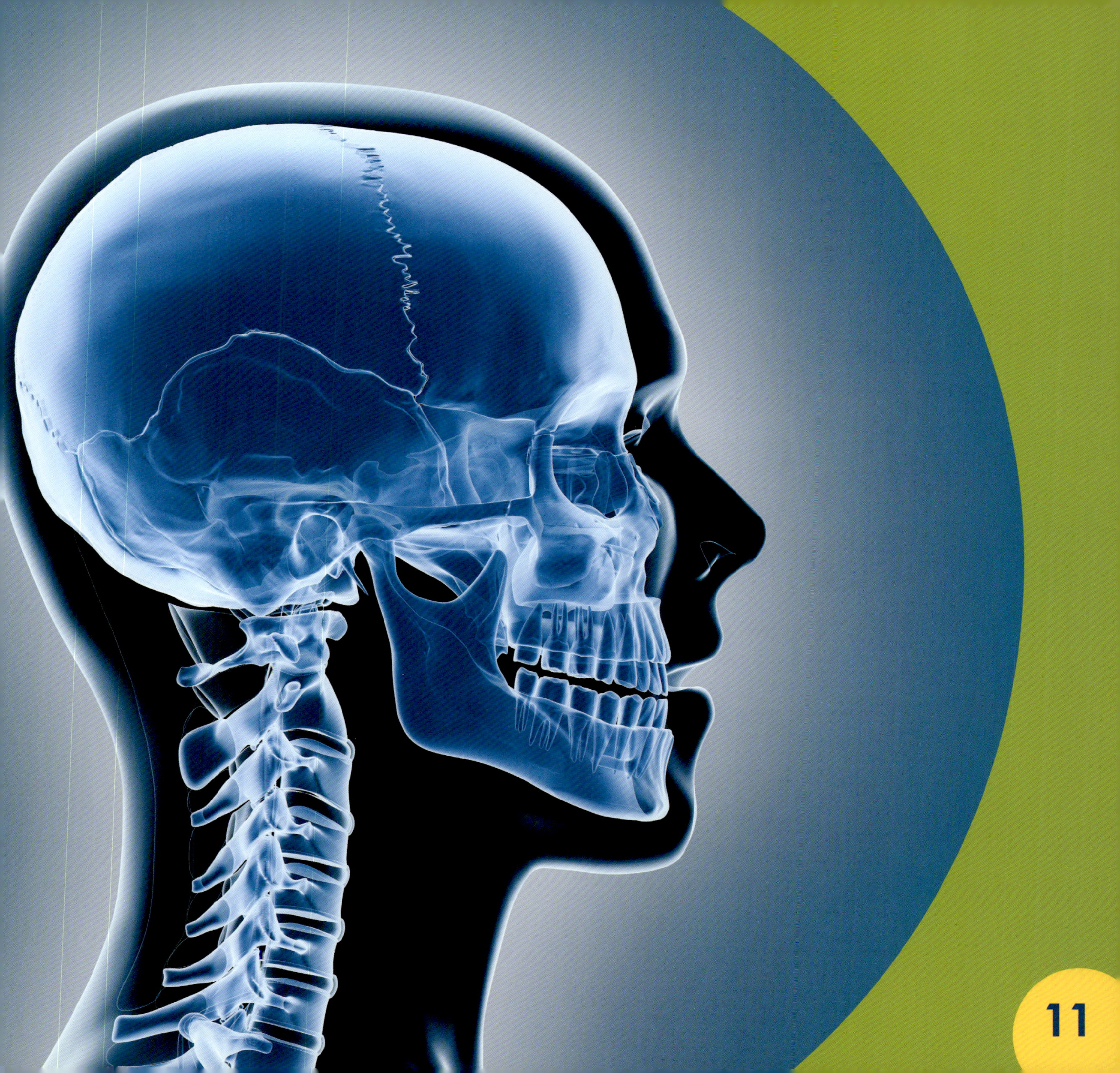

Spine

More than 90 percent of all animals do not have a spine. **The spine helps make humans unique.** It allows us to stand straight and walk on two legs.

The spine connects different parts of the skeleton. It is made of bones called vertebrae. These support the head, neck, and other body parts.

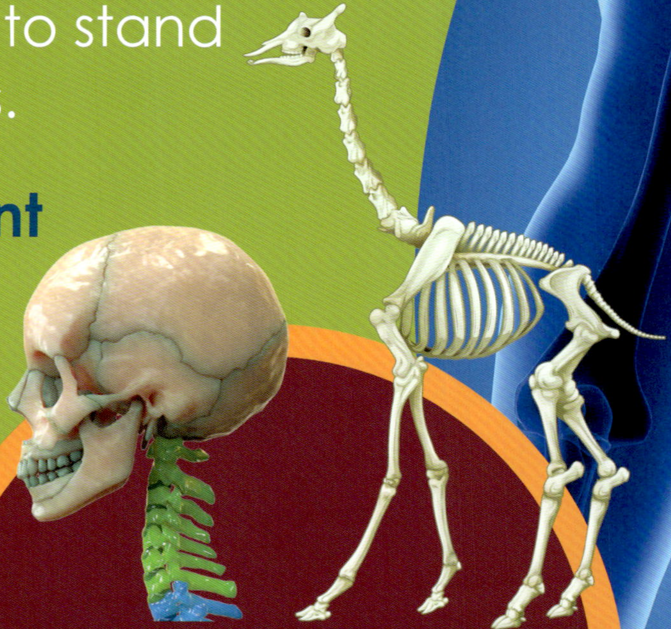

Both humans and giraffes have **seven vertebrae** in their necks.

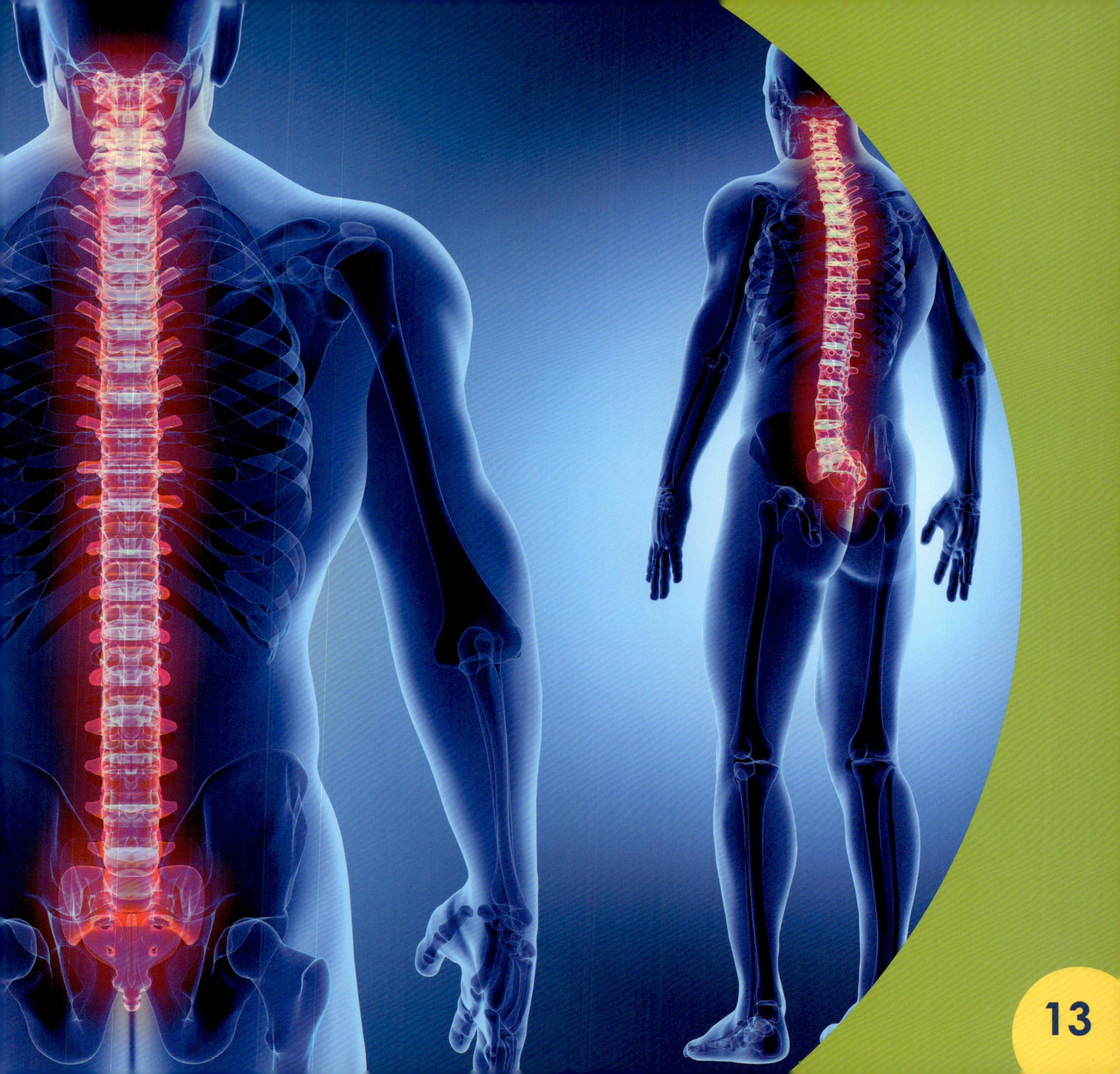

Arms and Hands

People use the bones in their arms and hands to type, write, and pick things up. **These bones are injured often.**

The wrist connects the arm to the hand. **Muscles in the hand move the fingers.**

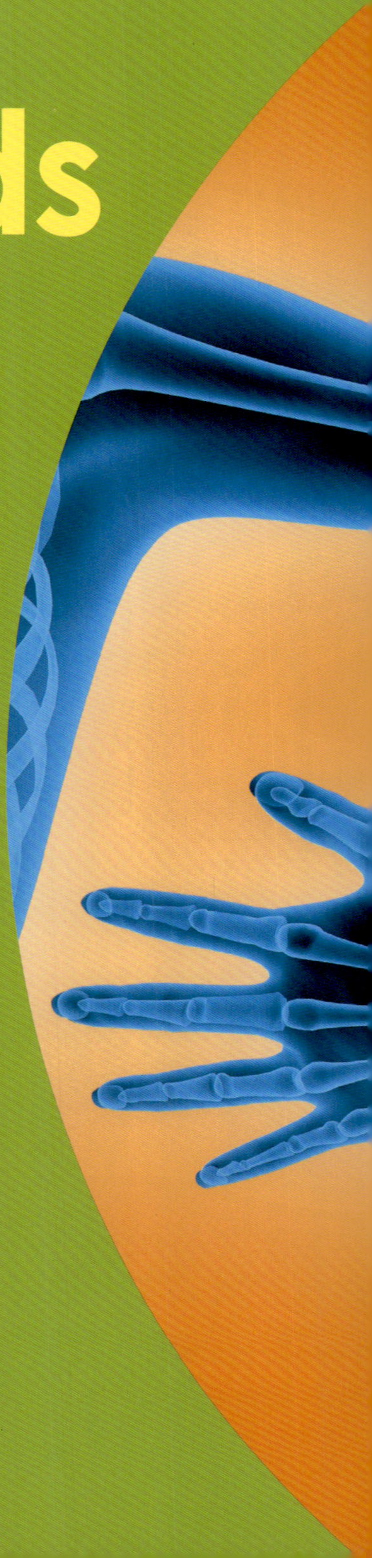

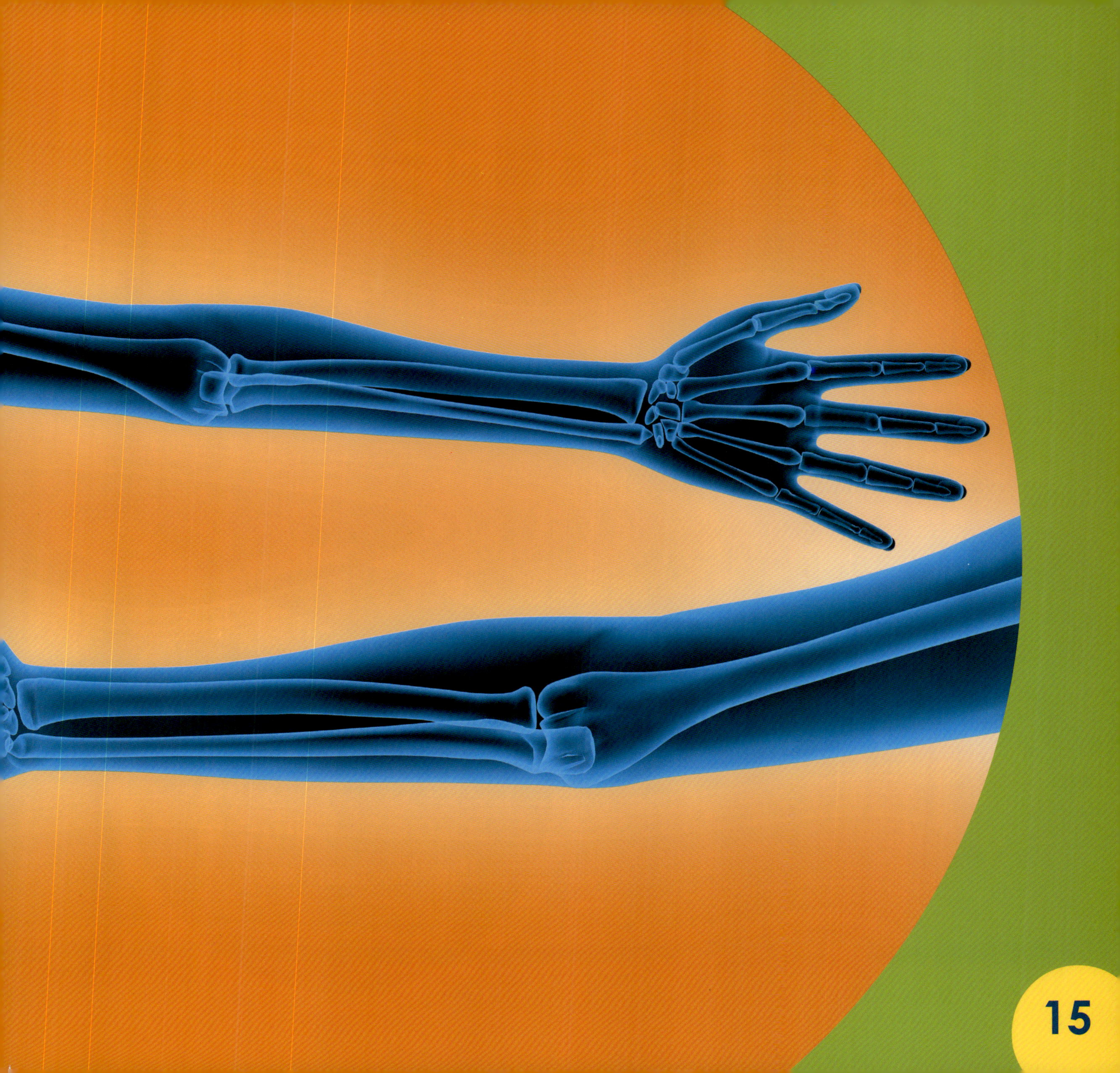

Legs and Feet

Legs and feet help people walk. **The bones in the legs are very strong.** They support the rest of the body. The femur in the upper leg is the strongest bone in the skeletal system.

The upper and lower legs are joined at the knee. The lower leg is connected to the foot.

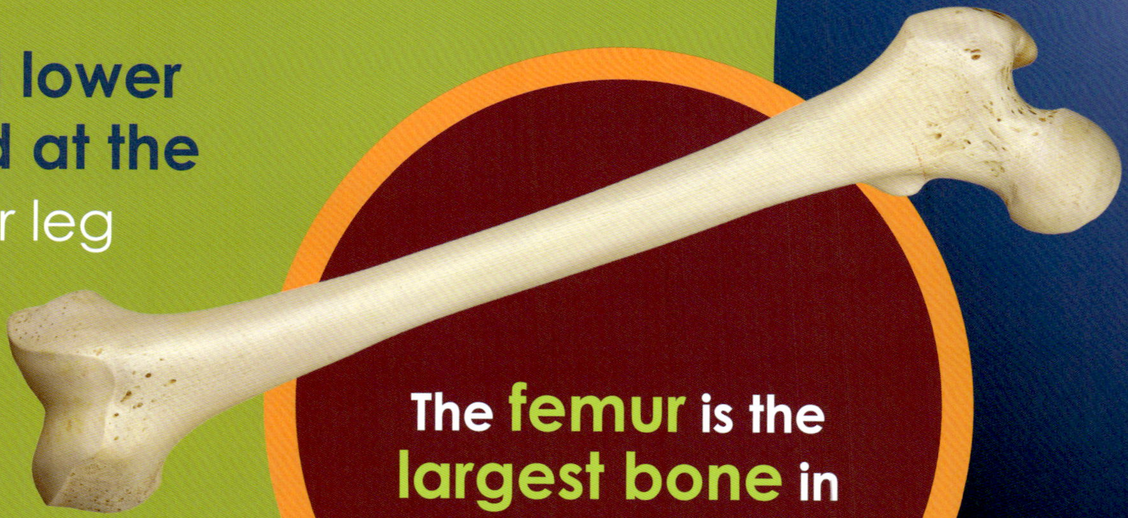

The **femur** is the **largest bone** in the human body.

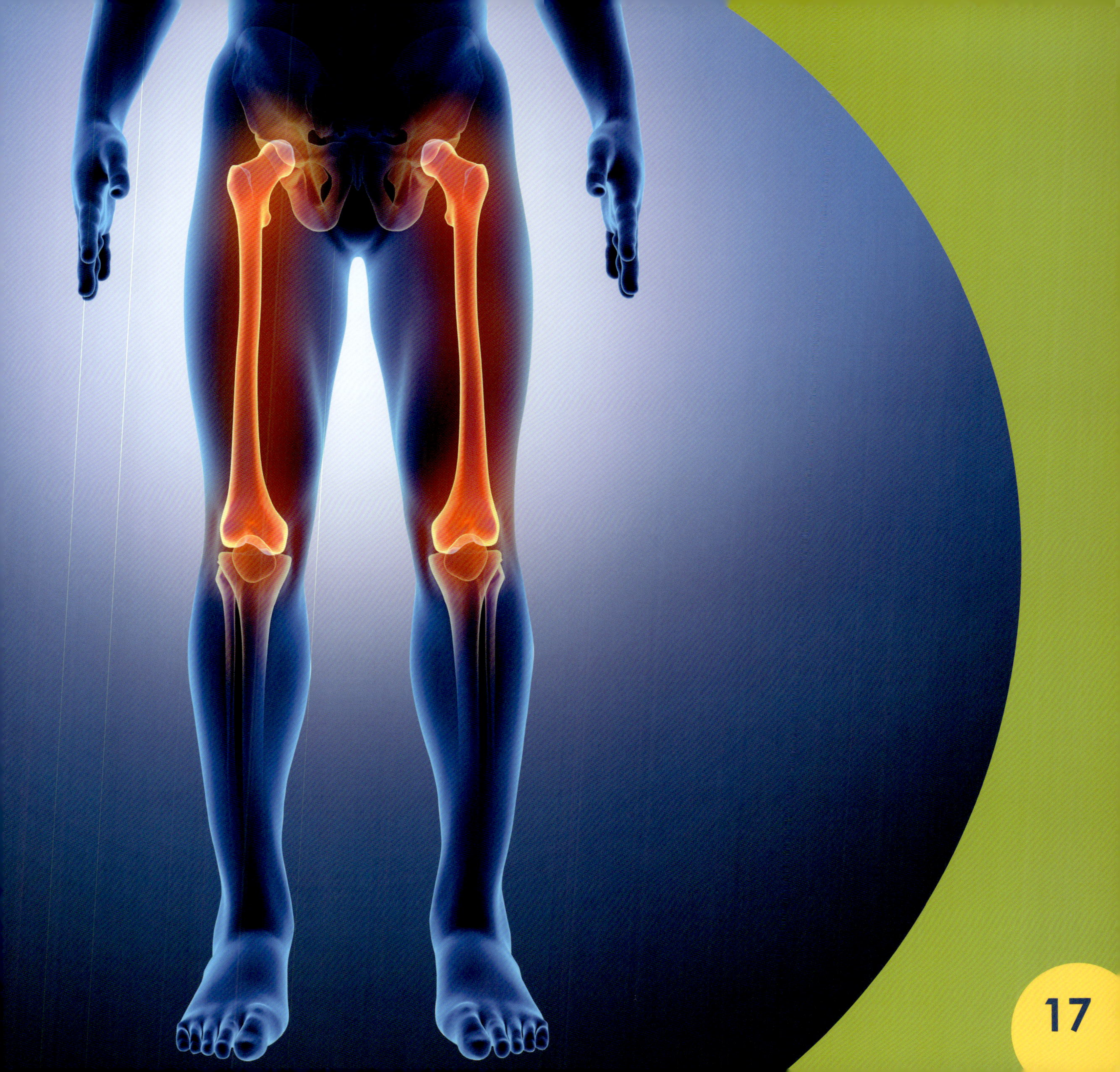

Staying Healthy

Did you know that bones need your help to stay healthy? **Calcium is good for bones.** Many foods are high in calcium.

Vitamin D helps the body absorb calcium. Sunlight gives the body vitamin D. So do some foods.

Career Spotlight

Physical therapists help people with injured or broken bones. They sometimes make casts. A cast holds broken bones in place until they heal. Physical therapists may suggest exercise plans to help bones heal faster.

New York City's Hospital for Special Surgery is one of the world's best hospitals for treating bone problems.

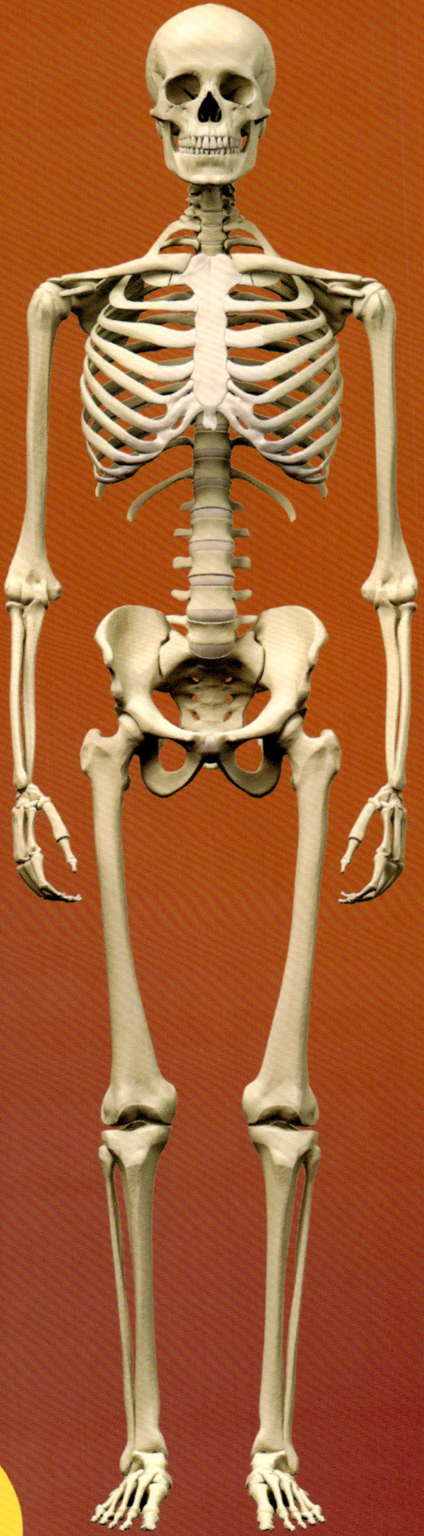

Skeletal System Quiz

Can you name the parts of the skeletal system shown in these pictures?

A] Skull

B] Spine

C] Arms and hands

D] Legs and feet

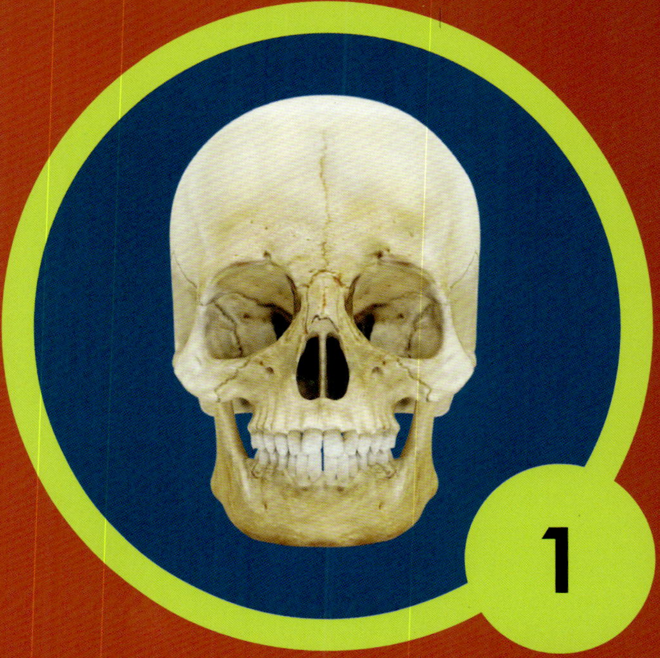

1

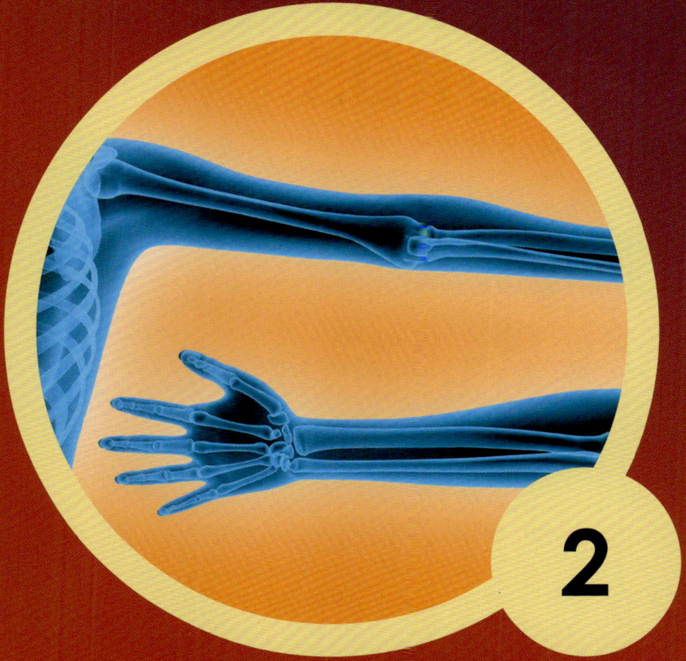

2

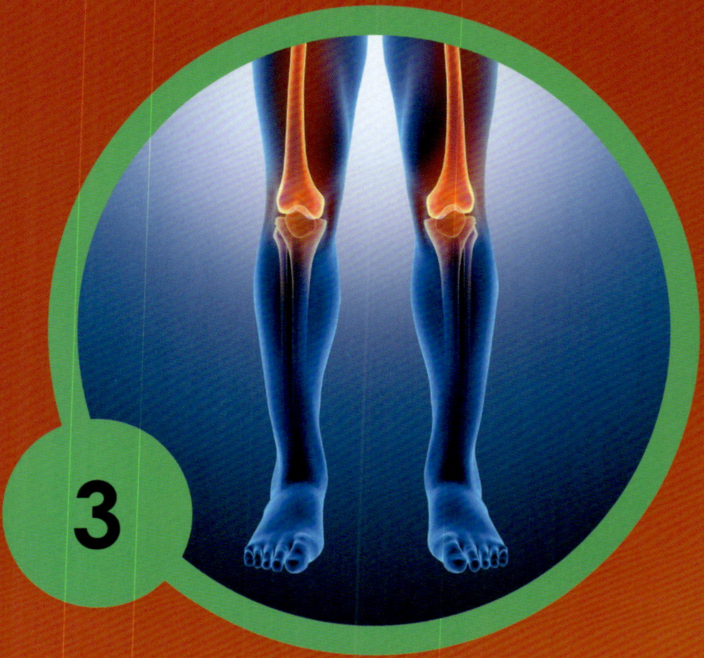

3

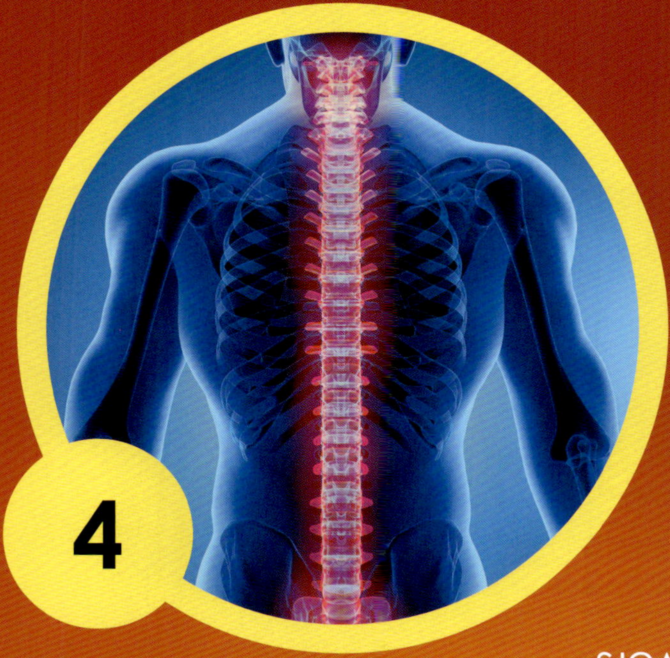

4

Answers
1] A
2] C
3] D
4] B

KEY WORDS

Research has shown that as much as 65 percent of all written material published in English is made up of 300 words. These 300 words cannot be taught using pictures or learned by sounding them out. They must be recognized by sight. This book contains 89 common sight words to help young readers improve their reading fluency and comprehension. This book also teaches young readers several important content words, such as proper nouns. These words are paired with pictures to aid in learning and improve understanding.

Page	Sight Words First Appearance
4	a, about, all, an, and, are, as, each, from, gives, important, is, it, its, like, made, of, such, than, the, they, think, times, your
6	can, children, few, grow, if, much, take, to
8	have, parts, these, together, with
10	at, do, eyes, face, in, move, not, one, only, other, there, this, too
12	animals, both, different, head, helps, make, more, on, two, us, walk
14	hands, often, people, their, things, up, use, write
16	feet, very
18	did, foods, for, good, high, know, many, need, that, you
19	so, some
20	may, or, place, sometimes, until, world

Page	Content Words First Appearance
4	body, bones, brain, harm, heart, job, machine, organs, shape, skeletal system, steel, systems
5	function, protection, support
6	adults, blood, oxygen, weeks
8	babies
9	carpal bones, femur, jawbone, leg, ribs, skull, spine, wrist
10	mouth, nose, teeth
12	giraffes, humans, legs, neck, vertebrae
14	arms, fingers, muscles
16	foot, knee
18	broccoli, calcium, cheese, milk, salmon, spinach
19	egg yolks, mushrooms, sunlight, vitamin D
20	casts, exercise plans, Hospital for Special Surgery, New York City, physical therapists, problems, spotlight

Published by AV2
14 Penn Plaza, 9th Floor New York, NY 10122
Website: www.av2books.com

Library of Congress Cataloging-in-Publication Data

Names: Das, Priyanka, author. | Kissock, Heather, author.
Title: Skeletal system / Priyanka Das and Heather Kissock.
Description: New York, NY : AV2 by Weigl, [2021] | Series: My first look at body systems | Audience: Grades 2-3 |
Identifiers: LCCN 2020018434 (print) | LCCN 2020018435 (ebook) | ISBN 9781791118761 (library binding) | ISBN 9781791118778 (paperback) | ISBN 9781791118785 | ISBN 9781791118792
Subjects: LCSH: Skeleton--Juvenile literature.
Classification: LCC QM101 .D37 2021 (print) | LCC QM101 (ebook) | DDC 611/.71--dc23
LC record available at https://lccn.loc.gov/2020018434
LC ebook record available at https://lccn.loc.gov/2020018435

Printed in Guangzhou, China
1 2 3 4 5 6 7 8 9 0 24 23 22 21 20

062020
100919

Project Coordinator: Priyanka Das Designer: Jean Faye Marie Rodriguez

Every reasonable effort has been made to trace ownership and to obtain permission to reprint copyright material. The publisher would be pleased to have any errors or omissions brought to its attention so that they may be corrected in subsequent printings.

The publisher acknowledges iStock and Shutterstock as the primary image suppliers for this title.